Florida Fossil Shark Teeth Identification Guide

The Fossil Shark Teeth Most Commonly Found In Florida

Florida Fossil Shark Teeth Identification Guide

The Fossil Shark Teeth Most Commonly Found In Florida

Robert L. Fuqua

Author's Cataloging In Publication Data
Fuqua, Robert L.
Florida Fossil Shark Teeth Identification Guide - The fossil shark teeth most commonly found in Florida
1. Fossils
2. Shark teeth
3. Florida
4. Paleontology
ISBN: 9798602488791

Table of Contents

Credits

All fossil shark teeth photographs are by the author

Shark jaw illustration on page 2 is by the author

Bull shark photograph on page 8 is by Albert Kok~enwiki and licensed under the Creative Commons Attribution-Share Alike 4.0 International license.

Great white shark photograph on page 10 is by Elias Levy and licensed under the Creative Commons Attribution 2.0 Generic license.

Hammerhead shark photograph on page 12 is by Berry Peters and licensed under the Creative Commons Attribution 2.0 Generic license.

Lemon shark photograph on page 14 is by Vivalatew and licensed under the Creative Commons Attribution-Share Alike 4.0 International license.

Mako shark photograph on page 16 is by Patrick Doll and licensed under the Creative Commons Attribution-Share Alike 3.0 Unported license.

Megalodon shark image on page 18 is by Dinosaur Zoo and licensed under the Creative Commons Attribution-Share Alike 3.0 Unported license.

Nurse shark photograph on page 22 is from NOAA and is in the public domain.

Sandbar shark photograph on page 24 is by Brian Gratwicke and licensed under the Creative Commons Attribution 2.0 Generic license.

Sand tiger shark photograph on page 26 is by UND77 and licensed under the Creative Commons Attribution-Share Alike 3.0 Unported license.

Snaggletooth shark photograph on page 28 is by Tassapon Krajangdara and licensed under the Creative Commons Attribution 3.0 Unported license.

Thresher shark tooth photograph on page 30 is by Thomas Alexander and licensed under the Creative Commons Attribution-Share Alike 4.0 International license.

Tiger shark tooth photograph on page 32 is by Albert Kok and licensed under the Creative Commons Attribution-Share Alike 3.0 Unported license.

Photograph of the author on page 34 is by Linda Donovan.

Other books by Robert L. Fuqua

A Living Jewel
- A Beginner's Guide To Saltwater Aquariums
Published 1995, Second Edition 2009

Hunting Fossil Shark Teeth In Venice, Florida
The Complete Guide: On The Beach, SCUBA, and Inland
Published 2011

Fossil Shark Teeth Of Venice, Florida
- The Paleogeology Behind How Shark Teeth, And Other Fossils, Ended Up In The Venice Area
Published 2017

Algae The Alligator
- The Alligator That Was Raised By Turtles
Published 2017

A Brief History Of The Earth
Formation Of The Earth, Plate Tectonics, Volcanoes, Supercontinents, Sea Levels, Ice Ages, Climate And Life
Published 2019

Overview Of Fossil Shark Teeth

Now that you have started collecting fossil shark teeth, you are probably getting interested in learning what kind of sharks the teeth came from and how long ago they lived. These are good questions and can lead to a lifetime of learning about fossils, sharks, paleontology, and even geology. This book will help start that process by discussing the anatomy of a shark's jaw, shark tooth anatomy, and finally the identification of the more common shark teeth found in Florida.

The megalodon tooth on the cover of this book is one that was given to me by my wife. It is shown actual size and has great coloration. As you can see, it is perfect for the cover of this book. That was an amazing gift! The megalodon tooth on the preceding page and on page 7 is one I found SCUBA diving off Venice, Florida. It is 5.75 inches long, bigger but not as colorful. It was my best find ever. I treasure both of those teeth. If you are new to hunting fossil shark teeth you may be interested in my book *Hunting Fossil Shark Teeth In Venice, Florida.* This book, also known as the "blue book," is a basic guide to hunting on the beach, diving in the Gulf using SCUBA, and at inland sites. If you are more advanced and want to understand why there are so many shark teeth, and other fossils, in this area, then you may be interested in my book *Fossil Shark Teeth Of Venice, Florida.* This book, also known as the "yellow book," covers the paleogeology that led to all those fossils being here.

Shark jaw anatomy
When we visualize a shark's jaw, we think of the sharp white teeth that are clearly visible in the top and bottom jaws. These are called the primary teeth and are the ones that do the biting and cutting. But those teeth are actually only a fraction of the teeth the shark has at any one time. As the shark bites its

prey and then struggles with it, both the shark and the prey will be turning and twisting, one to get away and the other to consume a meal. This puts lots of strain on the shark's teeth, especially when the tooth hits a bone, and some of them can get broken or lost in the struggle. Since a shark depends on a good set of sharp teeth for its survival, it can not afford to loose too many teeth. So, to survive the shark needs a constant supply of new, sharp, teeth to replace the ones that have broken or fallen out. You may find a tooth with a broken tip or a chunk missing from the side. This damage is generally caused by the shark biting into bone when it was feeding.

If you look at a shark's jaw from the inside perspective you will see numerous rows of teeth laying flat against the inside of the jaw, on both the top and bottom. The teeth nearest the active teeth are almost ready for use and the ones deepest inside are still developing. The active teeth are constantly being replaced by a process in which they continually move forward until they fall out of the mouth. At the same time, the next row of teeth in the jaw moves forward to replace the active teeth and the developing teeth also continually move forward. This process ensures that the shark is always well equipped with a full set of sharp teeth. Figure 1 shows the inside view of a shark jaw.

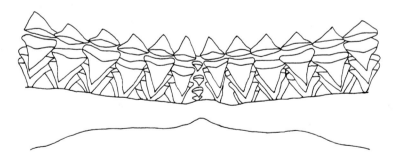

Figure 1. Inside view of a shark's jaw.

In addition to the rows of teeth, it is interesting to note the teeth that are sometimes found right in the midline where the left and right jaws join. These teeth are called symphyseal teeth, are considerably smaller than their companions, and usually have a significantly different shape. Four small symphyseal teeth can be seen in the center of Figure 1.

There is not much information available on how long the different species of sharks live. Some sharks live for about 16 years. Others can live for over 40 years. A shark can produce 2000 to 3000 teeth or more in a lifetime. When you consider how many sharks live today and how many have lived over the past 400 million years, there are a lot of shark teeth still waiting to be found worldwide.

Shark tooth anatomy

Not all teeth in a shark's jaw are shaped the same. Their teeth can be very different front to back and top to bottom. This difference can be in size, shape or both. The shape of the teeth also changes with age, whether it was an active or developing tooth, and sometimes based on the shark's sex. It is also true that the teeth from different species of sharks can be very similar. Because of this, it is sometimes difficult to correctly identify what kind of shark any given tooth came from. Another aspect of shark teeth to be aware of is their orientation in the shark's mouth. All of the teeth shown in this book are actual size unless otherwise noted.

The flatter side of the teeth face the front, outward side of the jaw. The side of the tooth that is oriented toward the inner mouth is usually a bit rounder and more attractive. This is the side that we generally display when showing off our finds. The flatter, front side slices into the prey and the curved back side starts forcing the flesh back into the mouth. This combination makes the tooth very efficient at removing flesh from the prey.

Beyond the size and shape of the teeth, there are other characteristics and descriptive terms that need to be understood.

3

These terms and characteristics are described below.

Blade or Crown - the exposed part of the tooth which is covered with enamel.

Bourlette - sometimes called the dental band - this is the dark chevron-shaped band that separates the blade from the root on some species.

Cusp - small blades that exist on either side of the main blade.

Cutting teeth - have broad, sharp blades that are efficient at removing flesh.

Distal side - this is the side of the tooth that points away from the center line of the jaw.

Grasping teeth - are narrow and pointed and used to restrain prey in the mouth.

Notch - is a distinct change in the shape of the side of the blade.

Nutrient groove - is a small groove often surrounding the nutrient pore.

Nutrient pore - is a small hole in the root through which blood and the nerves entered the tooth.

Root - is the base of the tooth where it is attached to the jaw.

Root lobe - refers to the sides of the root and is normally used to describe roots with pronounced ends.

Serrations - small, sharp notches along the side of the blade - used to increase the cutting ability of a tooth.

Shoulder - the area of the blade where it joins with the root.

Symphyseal teeth - small teeth that are sometimes found on the center line of the shark's jaws and often have a very different size and shape from those on the right or left sides of the jaw.

Other characteristics of fossil shark teeth

Most fossils in Florida are from the Miocene or Pliocene and are between 2-23 million years old. Some fossils found in the northern part of the state can be from the Oligocene or Eocene and can be up to 30-50 million years old. This is all explained in the "yellow book".

4

All fossils get their color based on the chemicals that were in sediment they rested in while mineralization was taking place. If there was lots of phosphate present in the sediment the fossil will be black. Iron in the sediment turns the fossil reddish orange. Limestone turns the fossil gray or yellow. Tannins turn the fossil dark brown. It is also possible for a fossil to have multiple colors either because of those chemicals being in the sediment or because the sediment itself changed over time. For example if the fossil was disturbed from its original resting place and was washed into another area.

When measuring a shark tooth, we measure from the point of the crown to the end of the root lobe. Since the measurement to each lobe will be different we use the longer of the two lengths when describing the tooth.

Perhaps most rare of all are the teeth that have natural pathological deformities such as dents, unusual twists, curves or splits that are the result of disease or healed injury that the shark suffered during its life.

You will find lots of teeth. Some will be in very good shape and others will be broken. Obviously an unbroken tooth is more desirable. So, we need to discuss what else makes one tooth better than another. Here are some characteristics that are important, in no order of relevance. A perfect larger tooth is generally more desirable than a perfect smaller tooth from the same kind of shark. It is also desirable that the tooth be intact with no chips or cracks. These teeth have been submerged in sediment, fossilized, and then possibly moved about on the land and/or under the water for millions of years. A lot can happen to a tooth during this time.

The blade surface of the tooth should be smooth and shiny on both the front and back. A tooth with scratch

5

marks or blemishes will be less desirable than one with perfect surfaces. If the tooth has serrations, they should all be present and sharp. Spending millions of years in the dirt or under water can cause damage to those sharp little serrations.

However, this does not mean that a less than perfect tooth is not valuable at all. When you are starting out, every tooth is a valuable find. Less than perfect teeth can also be attractively displayed in a jar with lots of other teeth or by embedding the broken part into a bone or rock to hide the missing part. Finally, imagine a kid out looking for teeth on the beach and not finding any. Giving that kid a handful of teeth, even broken ones, will help create a fun and memorable day.

Bull Shark

Upper teeth

Back side   Front side

Lower teeth

Figure 2. Bull shark and fossil bull shark teeth, *Carcharhinus leucas.*

Bull Shark

Bull sharks first emerged in the Miocene, still exist today, and probably have not changed much in the past 20 million years. Modern bull sharks can grow to 11 feet in length and weigh as much as 400 pounds. They are full-bodied sharks with a head that is more blunt than pointed. Bull sharks are found in warmer waters worldwide and prefer inshore and murky waters. They can also live in freshwater for extended periods of time and are often seen in rivers and lakes that connect with the ocean. Bull sharks are considered to be very dangerous to people.

Fossil bull shark teeth are very commonly found in Florida. See Figure 2. The upper teeth are about one inch long with a triangular blade that is slightly bent on both sides as it sweeps to join the root. The root is also curved. Front teeth are symmetrical but the teeth on the sides of the mouth sweep backwards. The tooth has fine serrations starting near the point of the blade and getting slightly more coarse as they approach the root, making it good for cutting.

Lower bull shark teeth are smaller than the uppers and much more peg like. They are stout and less broad than the upper teeth and are good for clutching prey. The root also has a curve and the overall root is quite wide when compared to the length of the blade. The lower teeth have very fine serrations that are sometimes hard to see, but can be felt with a fingernail.

Bull shark teeth are actually very similar to the teeth of several other members of the Carcharhinus genus such as the dusky shark. It is sometimes difficult to determine exactly which shark a tooth came from, especially when the differences between upper and lower jaws as well as front and back teeth are considered.

Great White Shark

Upper teeth

Back side   Front side

Lower teeth

Figure 3. Great white shark and fossil great white shark teeth, *Carcharodon carcharias*.

Great White Shark

Great white sharks are generally considered to have emerged in the early Miocene and still exist today. They are large, heavily bodied sharks that can reach over 20 feet in length, weigh as much as 5000 pounds, and have probably not changed much since the Miocene. Great white sharks have a gray body and white belly with one large and one small dorsal fin. They also have a sharp, pointed head. Great white sharks can be found worldwide but seem to prefer cool to warm waters so they are not commonly seen in equatorial or polar waters. They are a very dangerous shark and their bite can do massive damage to its prey.

Fossil teeth from great white sharks are sometimes found in Florida, however they are not as common as bull shark teeth. See Figure 3. Teeth from the upper jaw of the great white shark can reach over two inches in length. They have a broad, flat root giving the overall tooth a triangular shape. The blade of the tooth is fully serrated with coarse serrations.

The lower teeth are somewhat smaller and the sides of the blade usually curve inward a small amount rather than straighter like the upper teeth. The blade of the tooth is also thicker where it connects to the root. They are obviously good at clutching prey while the upper teeth do the cutting.

Great white shark teeth are often compared to the teeth from the *Carcharocles megalodon* (which will be discussed later). There are several distinctions that allow us to tell them apart. Megalodon teeth are typically much larger and have much finer serrations. Megalodon teeth also have a bourlette between the blade and the root.

11

Hammerhead Shark

Upper teeth

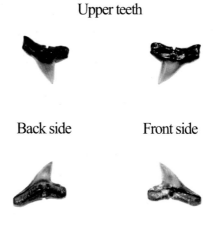

Back side   Front side

Lower teeth

Figure 4. Hammerhead shark and fossil hammerhead shark teeth *Sphyrna laevissima.*

Hammerhead Shark

Hammerhead sharks first emerged in the Miocene and eventually evolved into the modern species that still exists today. The fossil hammerhead teeth that we find belong to the species *Sphyrna laevissima,* which is extinct. Hammerhead sharks are probably the most distinctive shark because of their unusually shaped head. The head has evolved into two flat lobes that extend out from the center. These lobes contain the eyes and other sensory organs. The mouth is located below the center of the head as in any other shark. They can reach up to 20 feet in length and have gray bodies and white bellies. The modern hammerheads probably look very similar to the extinct *Sphyrna laevissima.* Hammerheads are found worldwide in more tropical waters. Some, but not all, modern hammerheads are dangerous to humans.

Fossil teeth from hammerhead sharks are commonly found in Florida. They are found more often than great white shark teeth but not nearly as often as bull shark teeth. The teeth of the hammerhead are surprising in that they are much smaller than expected for a shark that can grow so large. They are about half an inch long. See Figure 4. The distal side of the blade has a distinct notch giving the tooth a very angular shape. Where this notch approaches the root, the area referred to as the shoulder, there is a thin section of the blade which is sometimes serrated. Hammerhead teeth also have a very deep and distinct nutrient groove which can be easily seen in the center of the root.

The lower teeth are very similar in size and shape to the uppers. Both the upper and lower teeth have a good combination of clutching and cutting capabilities.

13

Lemon Shark

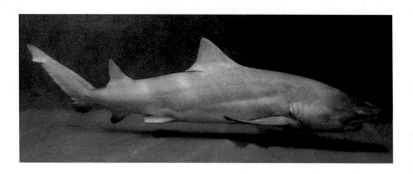

Upper teeth

Back side   Front side

Lower teeth

Figure 5. Lemon shark and fossil lemon shark teeth, *Negaprion eurybathrodon.*

Lemon Shark

Lemon sharks first emerged in the Miocene and are still in existence today. Lemon sharks can grow up to 13 feet in length and can weigh up to 400 pounds. They have two large dorsal fins and get their name from the pale yellowish brown coloring of their bodies. They prefer warmer waters and are found in the southern Atlantic Ocean, in the Caribbean, along the eastern United States, in the Gulf of Mexico, and off the west coast of Mexico and Central America. Even though they are a large shark, they are not generally considered to be dangerous to humans.

Fossil teeth from lemon sharks are very readily found in Florida and are just as common as bull shark teeth. See Figure 5. Upper lemon shark teeth can be up to one inch long and the blade is somewhat narrow and long making it a good combination clutching and cutting tooth. The teeth in the front of the mouth stand upright and look like a "T," while the rest have a slight angle to them. One characteristic of upper lemon shark teeth is that the blade has thin shoulders that spread out over the top of the root. Sometimes these shoulders have slight serrations. The roots are broad with a flat to slightly curved base.

The lower teeth are shaped very similarly to the uppers and also tend to stand perpendicular to the base or have a slight bend. However, the blade is less broad which results in a good clutching tooth. Also, the shoulders where the blade meets the root are much thinner than on the upper teeth. In some cases, the shoulders can barely be seen.

15

Mako Shark

Upper teeth

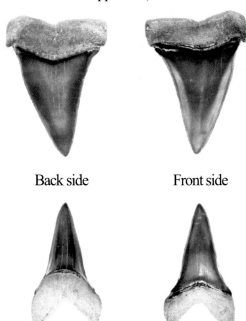

Back side   Front side

Lower teeth

Figure 6. Mako shark and fossil mako shark teeth., *Isurus hastalis.*

Mako Sharks

Mako sharks emerged in the Miocene, and while some species became extinct, others survived and still exist today. The extinct species *Isurus hastalis* is the most commonly found fossil mako shark tooth in Florida. Some species of modern mako sharks can grow to 13 feet long and weigh over 2,000 pounds. They have one large dorsal fin and a second one that is very small. The tail is somewhat crescent shaped. Their bodies are bluish and they have white bellies. Modern mako sharks probably look very similar to their extinct cousins. These sharks can be found worldwide but prefer warmer waters, so they are usually found in the mid latitudes. Some mako sharks are considered to be dangerous to humans.

Fossil teeth from mako sharks are often found in Florida, but they are not as common as either bull or lemon shark teeth. See Figure 6. The mako shark's upper teeth are flat, broad triangles that can reach over two inches in length. They are very similar in size and shape to great white shark teeth, except that mako teeth do not have serrations. The mako tooth can also be canted at a shallow angle.

The lower teeth have a very different shape. They are long and pointed without any serrations. These teeth are rather stout, especially where the blade meets the root. They are obviously good at clutching big prey while the upper teeth do the cutting.

17

Megalodon

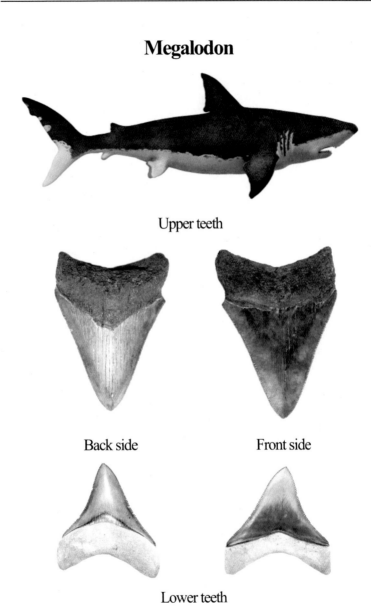

Upper teeth

Back side   Front side

Lower teeth

Figure 7. Megalodon and fossil megalodon teeth, *Carcharocles megalodon.* The teeth photographs are half their actual size.

Megalodon

Megalodons are by far the most impressive sharks that have ever existed. They lived from 30 to 2 million years ago, so they have been extinct for a long time. Megalodons reached lengths of 70 feet with a mouth approximately six feet wide and six feet high. These giant sharks ranged worldwide in warmer waters and fed on whales and other large marine mammals. Megalodons are often referred to by the shortened name "Meg." Giant sharks have to have giant teeth and the megalodon's can be up to seven inches measured from the point of the blade to the farthest edge of the root. The size of a tooth corresponds to the size of the megalodon by a ratio of about ten feet of megalodon length to each inch of tooth.

Megalodon teeth can be found in Florida but you have to work hard to find one. See Figure 7. Upper megalodon teeth are not only long, they are also very broad and heavy. They have a good, solid feel to them. Megalodon teeth have fine serrations along the blade on both sides. Another distinguishing feature is the bourlette, or dental band, which is the dark chevron-shaped band that separates the blade from the root on the back side of the tooth. There are some similarities between smaller megalodon teeth and large great white teeth. The way to tell the difference is that great white teeth do not have this dark band. Great white teeth also have larger, much coarser serrations.

Lower megalodon teeth are less broad and have more of a curve on the sides of the blade. They are also much thicker where the blade joins the root. Lower megalodon teeth also have fine serrations and a bourlette.

Megalodon Ancestors
Back side   All upper teeth   Front side

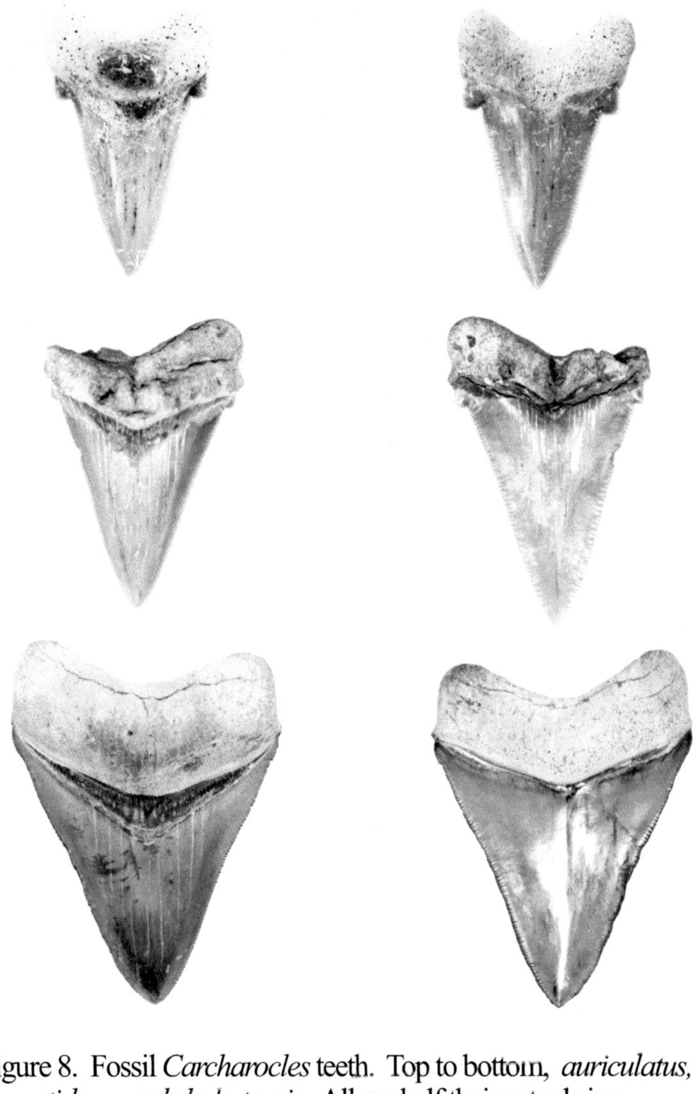

Figure 8. Fossil *Carcharocles* teeth. Top to bottom, *auriculatus, augustidens, and chubutensis.* All are half their actual size.

Megalodon Ancestors

Megalodons existed from 30 to 2 million years ago. They were the end of the line of the Carcharocles genus. Teeth from three of its ancestors can be found in Florida, however they are even more rare than megalodon teeth. Mostly they are found in northern Florida which is older than southern Florida. This is explained in my yellow book, *Fossil Shark Teeth Of Venice, Florida.*

The oldest of these three ancestors is the *Carchorocles auriculatus* which existed from 35 to 25 million years ago. The *auriculatus* was about 30 to 35 feet long and had a narrow blade with serrated teeth. They were up to 4.5 inches long. The notable thing about these teeth is that they have pronounced and slightly rounded cusps on each side of the base of the blade.

Next came the *Carcharocles augustidens* which existed from 33 to 22 million years ago. The *augustidens* was about 30-35 feet long and also had narrow, serrated teeth up to 4.5 inches long. They also had cusps, but they were smaller than those of the *auriculatus*.

The most recent ancestor of the megalodon is the *Carcharocles chubutensis* which existed from 28 to 5 million years ago and were about 50 feet long. Their teeth were wide and very similar to those of the megalodon except that they were only up to five inches long and had vestigial cusps. These vestigial cusps are sometime subtle and easily overlooked. Look at the lower left tooth on the previous page. A small vestigial cusp can be seen on the left side at the base of the blade. These cusps are totally absent in the megalodon. So, we can see how the tooth evolved from long, narrow teeth with pronounced cusps to longer, wider teeth with no cusps.

21

Nurse Shark

Upper teeth

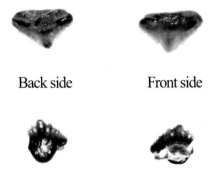

Back side   Front side

Lower teeth

Figure 9. Nurse shark and fossil nurse shark teeth, *Ginglymostoma lehneri.*

Nurse Shark

Nurse sharks came into existence 110 million years ago during the Cretaceous. They can reach up to 10 feet in length and weigh up to 250 pounds. Nurse sharks have a rounded head that is somewhat flattened. They also have two large dorsal fins and a large tail. Their color is a brownish gray. Nurse sharks are found in the warmer, coastal waters on both sides of the Atlantic as well as the eastern Pacific. They are bottom dwellers that have two barbels on the upper jaw. These barbels are used to detect prey buried in the ocean floor. Nurse sharks are suction feeders which means that they feed by rapidly sucking water into their mouths pulling in their prey. The prey is then ground into smaller pieces for swallowing. A friend of mine was bit on the side of his head by a nurse shark and he said the suction was so strong it felt like it was going to rupture his ear drum.

Fossil nurse shark teeth are rarely found in Florida. Those that are found come from the Peace River and up into northern Florida. The biggest teeth are about one half an inch long, but many of them are very small, less than .25 inch. See Figure 9. The small sizes contribute to them being hard to find.

Nurse shark teeth have a very unusual shape. They look more like the rear teeth of a dog than a typical shark tooth. These teeth have a central point with several cusps on each side. Upper and lower teeth are similar in size and shape.

Sandbar Shark

Upper teeth

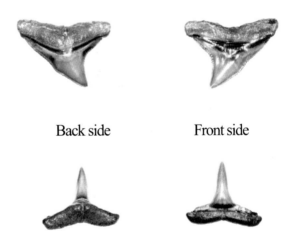

Back side   Front side

Lower teeth

Figure 10. Sandbar shark and fossil sandbar shark teeth, *Carcharhinus plumbeus*.

Sandbar Shark

Sandbar sharks came into existence during the Miocene and still exist today. They can reach 6-8 feet in length and weigh up to 200 pounds. Sandbar sharks have heavy bodies and rounded snouts and look very similar to other Carcharhinus sharks like the bull and dusky. They have one large and one small dorsal fin and are grey brown to bronze on their back and sides. They are white underneath. These sharks are not considered to be dangerous to humans.

Sandbar sharks can be found in warmer coastal waters worldwide, but are known to go deeper during seasonal migrations to warmer waters.

Fossil sandbar shark teeth are commonly found in Florida but slightly less often than bull shark teeth. Their teeth are very similar in size to bull shark but a bit thinner and have a more pronounced curve. See Figure 10. The upper teeth are about one inch long with a triangular blade that is slightly bent on both sides as it sweeps to join the root. The root is also curved. Front teeth are symmetrical but the teeth on the sides of the mouth sweep backwards. The tooth has fine serrations starting near the point of the blade and getting slightly more coarse as they approach the root, making it good for cutting. The main difference between sandbar shark teeth and other Carcharhinus teeth is that sandbar teeth are a bit thinner.

Lower sandbar shark teeth are smaller than the uppers and much more peg like. They have good clutching blades that are stout and less broad than the upper teeth. The root is quite wide when compared to the length of the blade. The lower teeth have very fine serrations that are sometimes hard to see, but can be felt with a fingernail.

25

Sand Tiger Shark

Upper teeth

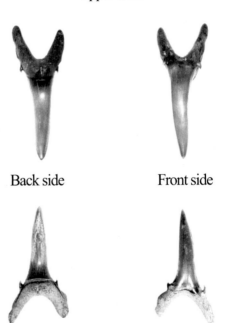

Back side   Front side

Lower teeth

Figure 11. Sand tiger shark and fossil sand tiger shark teeth, *Carcharias cuspidate.*

Sand Tiger Shark

The sand tiger shark first emerged in the Miocene, and while some species became extinct, others survived and still exist today. Modern sand tiger sharks probably look very similar to their extinct cousins.

Sand tiger sharks can grow to ten feet and weigh up to 350 pounds. They have a gray back and white belly with two large dorsal fins. Sand tiger sharks have a back that appears somewhat humped and a very flat head giving it an almost shovel-nosed appearance. But, perhaps their most notable characteristic is the mouth full of long, pointed teeth that look particularly menacing. These sharks can be found worldwide in warmer waters. Sand tiger sharks have been known to attack humans.

Fossil sand tiger shark teeth are commonly found in Florida. The species *Carcharias cuspidate* is the most commonly found fossil sand tiger tooth. See Figure 11. Sand tiger shark teeth can reach over one-and-a-half inches long. They have a very dramatic appearance starting with a deep "U" shaped root and a long pointed blade with a notable front to back curve. Many of these teeth also have a small cusp on each side of the blade.

The lower teeth are very similar in appearance. All of these teeth are very good for clutching and holding their prey.

Snaggletooth Shark

Upper teeth

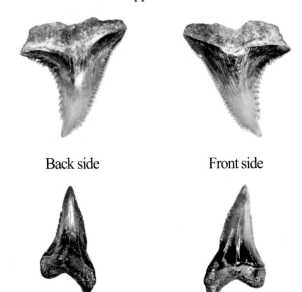

Back side   Front side

Lower teeth

Figure 12. Snaggletooth shark and fossil snaggletooth shark teeth, *Hemipristis serra.*

Snaggletooth Shark

The snaggletooth shark first emerged in the Miocene and eventually evolved into the modern species that still exists today. The fossil snaggletooth teeth that we find belong to the *Hemipristis serra* species which is extinct. Even though the proper name of this shark is the snaggletooth, fossil collectors generally refer to it as the "hemi," which is a shortened version of the shark's genus, Hemipristis. Modern snaggletooth sharks grow to about eight feet in length. Based on tooth size, the extinct snaggletooth shark was probably a bit larger. These modern sharks are found in the Indian ocean and western Pacific. The extinct snaggletooth shark lived in what is now the Atlantic Ocean. These sharks are not considered to be very dangerous to humans.

Fossil teeth from the snaggletooth shark are often found in Florida and are about as common as tiger shark teeth, but not as common as bull shark teeth. See Figure 12. The upper teeth of the snaggletooth shark are quite distinct and highly prized by fossil shark tooth collectors. They can reach up to and over one-and-a-half inches long, have a large, broad, curved blade with heavy serrations. These teeth are obviously very efficient at slicing through the shark's prey. The root is small by comparison and has a small peak in the middle of the back side of the tooth.

The lower teeth have a very different shape. They are also up to one-and-a-half inches long and pointed. These teeth are rather stout and were obviously good at clutching prey while the upper teeth did the cutting. One distinguishing characteristic of the lower tooth is a very bulbous protuberance where the middle of the blade joins the root.

Thresher Shark

Upper teeth

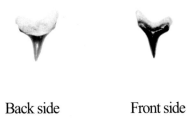

Back side   Front side

Lower teeth

Figure 13. Thresher shark and fossil thresher shark teeth, *Alopias superciliosus.*

Thresher Shark

The thresher shark emerged around 49 million years ago. They grow up to 16 feet long and can weigh up to 1000 pounds. Thresher sharks are brownish to blue-gray on top with a white undersides. They have a short head with big eyes and a pointed conical nose. Threshers only have one rounded dorsal fin. Their most distinctive feature is their very long slender tail which is used to corral prey and force them to swim toward the mouth. It can also be used to stun prey.

They can be found in warmer waters but are more pelagic preferring deeper waters than coastal waters. Thresher sharks are not considered to be dangerous to humans.

Thresher shark teeth, *Alopias superciliosus* can be found in Florida but are not very common. See Figure 13. The upper teeth are usually less than inch and long and have a stout appearance. The root can have a broad smooth curve. Upper teeth are smooth and often have a nice curve.

Lower thresher shark teeth also have a stout appearance but are a bit smaller than the uppers.

Tiger Shark

Upper teeth

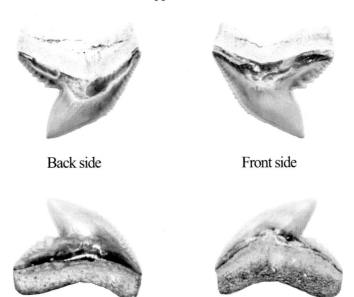

Back side   Front side

Lower teeth

Figure 14. Tiger shark and fossil tiger shark teeth, *Galeocerdo species.*

Tiger Shark

Tiger shark first emerged in the Miocene and eventually evolved into the modern species that still exists today. The fossil tiger shark teeth that we find belong to several species, (*Galeocerdo species*) some of which are extinct.

Modern tiger sharks are big and can grow to over 15 feet in length and can weigh over 1,000 pounds. These sharks have gray bodies and white bellies with one large and one small dorsal fin and a tall caudal fin. The sides of their bodies are marked with dark spots and vertical bars which is how they got their common name. These sharks are considered to be very dangerous to humans.

Fossil teeth from the tiger shark are often found in Florida and are about as common as hemi teeth, but not as common as bull shark teeth. See Figure 14. The teeth of the tiger shark are quite distinct and highly prized by fossil shark tooth collectors. They can reach over one-and-a-quarter inches long and have a large, broad, curved blade with a very distinctive notch. These teeth have coarse serrations on the shoulders of the blade and finer to non-existent serrations toward the point of the blade.

The lower teeth are very similar in size and shape to the upper teeth. These teeth are obviously very efficient at both clutching and slicing through the shark's prey. Tiger sharks like to eat turtles and their teeth are very able to cut through a turtle shell.

33

Robert L. Fuqua

About The Author

Robert L. (Bob) Fuqua was born in south-central Kansas in October 1949 and grew up steeped in mid-western values and work ethic. In 1971, he graduated from Kansas State University with a degree in Mechanical Engineering and then served in the U.S. Air Force for four years doing intelligence work. That led to a civilian career in intelligence in Maryland. In 2002, he retired with over 30 years service. His wife Linda had previously retired, also with over 30 years service in intelligence. They moved to Florida in 2004.

Bob started keeping freshwater aquariums as a young boy and was always interested in everything that lived in the water. He was also fascinated by the fact that Kansas was once part of a large inland sea and fondly remembers finding fossil shark teeth in the chalk bluffs of western Kansas when he was a young boy.

In 1977, he became a certified SCUBA diver and started making trips to the Florida Keys and Caribbean to dive on the coral reefs. He also dove on ship wrecks from New Jersey to Florida. In 1985, he volunteered as a SCUBA diver at the new National Aquarium in Baltimore and eventually became the Sunday Dive Captain. After five years, he had to give up that position to take a field assignment in Hawaii with his real job.

Bob's other interests include astronomy, bicycle riding, stand-up paddle boarding, and British sports cars. However, one of his greatest passions is diving for fossil shark teeth in the Gulf of Mexico off of Venice, Florida. He now has over 15,000 fossil shark teeth in his collection, including some very large megalodon teeth and those that he found in Kansas over 60 years ago.

Made in the USA
Monee, IL
18 November 2024

70097167R00026